cience all around me

Forces

Karen Bryant-Mole

RIGBY
INTERACTIVE
LIBRARY

This edition © 1997 Rigby Education
Published by Rigby Interactive Library,
an imprint of Rigby Education,
division of Reed Elsevier, Inc.
Chicago, Illinois

Customer Service 888-454-2279

Visit our website at www.heinemannlibrary.com

Printed and bound in Hong Kong
Cover designed by Lisa Buckley
Interior designed by Jean Wheeler
Commissioned photography by Sharon Hoogstraten and Zul Mukhida

06 05 04 03 02
10 9 8 7 6 5 4

Library of Congress Cataloging-in-Publication Data
Bryant-Mole, Karen.
    Forces / Karen Bryant-Mole
      p. cm. -- (Science all around me)
    Includes bibliographical references and index.
    Summary: Explains the basic principles of forces and movement through direct observation and through looking at everyday experiences.
    ISBN 1-57572-108-2 (lib. bdg.)   ISBN 1-4034-0050-4 (pbk. bdg.)
     1. Force and energy—Juvenile literature. [1. Force and energy.]
    I. Title. II. Series.
QC73.43B79 1996
531—dc20   96-22978
                                                        CIP
                                                        AC

Acknowledgments
The Publishers would like to thank the following for permission to reproduce photographs: Chapel Studios, p. 20; John Heinrich/Eye Ubiquitous, pp. 6, 8; Tony Stone Images, p. 16; Mike Powell, p. 18; Lori Adamski Peek, p. 22; Andrew Sacks/Zefa, pp. 4, 10, 12, 14.

Every effort has been made to contact copyright holders of any material reproduced in this book. Any omissions will be rectified in subsequent printings if notice is given to the Publisher.

Words that appear in the text in **bold** can be found in the glossary.

Contents

What Are Forces?

Forces are pushes and pulls.
Forces can make things move.

This bulldozer is acting as a force.
It is pushing the soil into a big pile.

(i) *Anything that starts moving is being pushed or pulled by something or someone.*

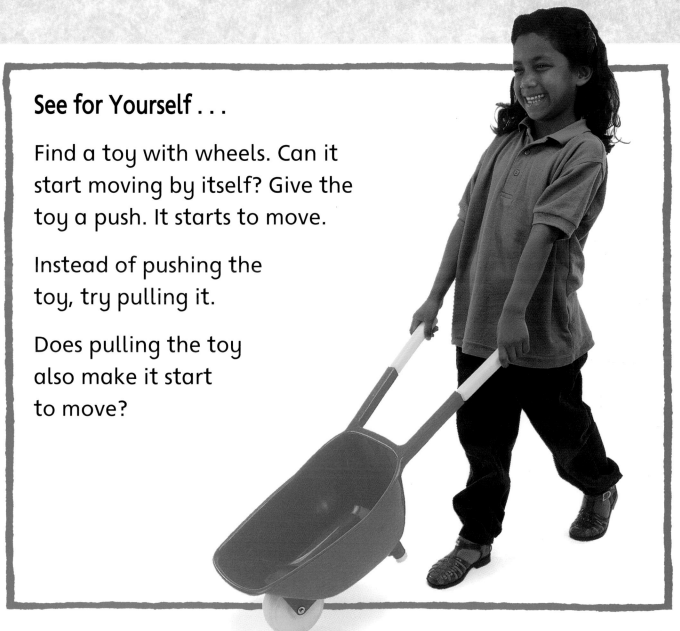

See for Yourself . . .

Find a toy with wheels. Can it start moving by itself? Give the toy a push. It starts to move.

Instead of pushing the toy, try pulling it.

Does pulling the toy also make it start to move?

5

Forces in Nature

Forces are found in the **natural** world all around us.

The wind is pushing this tree and making it bend.

Young plants push soil out of the way as they come up through the ground.

People are part of nature, too. We can push and pull with our bodies.

? *Which part of your body would you use to throw a ball?*

See for Yourself . . .

You can use your breath to push a ball. How?

Place a small ball inside a shoebox lid. Blow through the straw, using the push of the air to get the ball to the other end. You can turn this experiment into a game. Can you think how?

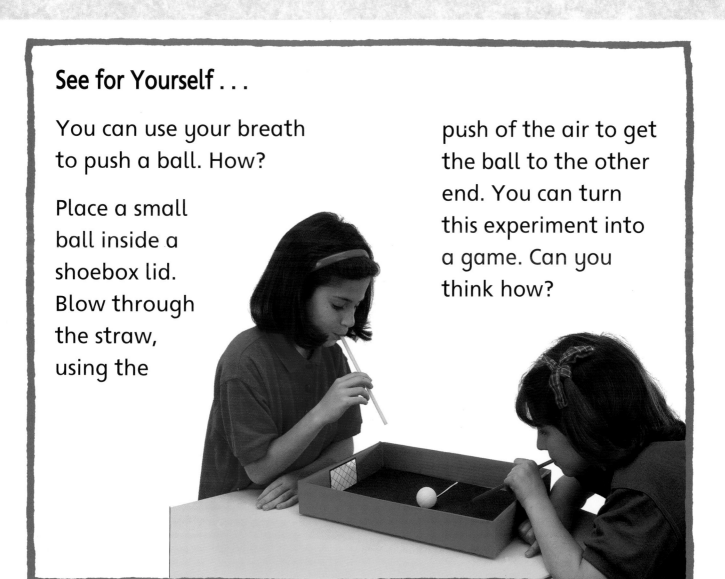

Manufactured Forces

Pushes and pulls can also be made,
or manufactured, by machines.

The front section of a train is called a *locomotive*.
This locomotive is powered by an engine that
pushes the locomotive along the track.

The carriages are
moving, too. They
are being pulled
along the track by
the locomotive.

? *What other vehicles need engines to make them move?*

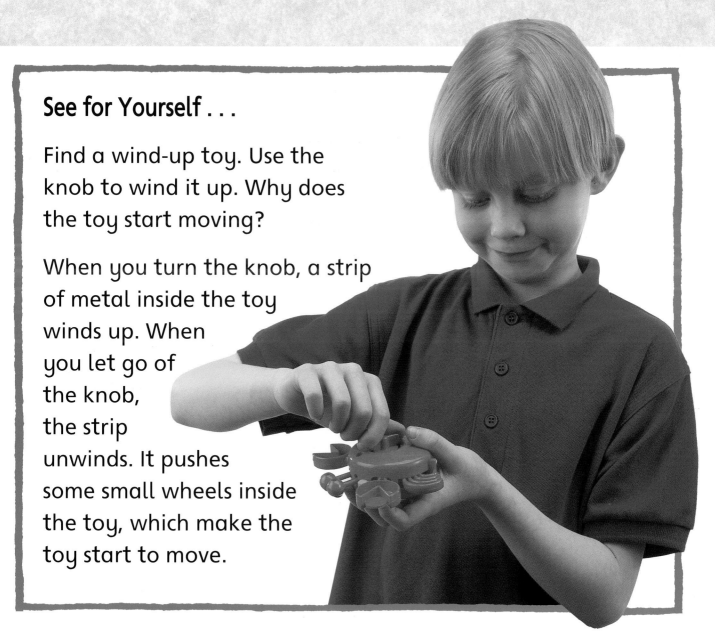

See for Yourself . . .

Find a wind-up toy. Use the knob to wind it up. Why does the toy start moving?

When you turn the knob, a strip of metal inside the toy winds up. When you let go of the knob, the strip unwinds. It pushes some small wheels inside the toy, which make the toy start to move.

Movement

As these racing cars **travel** around the race track, they will slow down, speed up, and turn around corners. They have started, and they will stop.

Movements like these are all caused by forces. Different types of force produce different types of movement.

? *The cars are moving forward. What is the opposite of forward?*

See for Yourself . . .

Play "Follow the Leader" with some friends.

Have one person be the caller, and another, the leader.

The caller shouts, "start," "stop," "faster," "slower," "backward," "forward," "turn left," or "turn right."

Everyone has to follow the leader.

Speed

These athletes are lining up to run a race. The fastest person will be the winner.

People run by pushing against the ground with their legs and feet.

Runners build strong **muscles** in their legs so that they can push hard and run fast.

? Can you feel your legs and feet push when you run?

See for Yourself . . .

Place a book on a smooth floor. Give it a gentle push. What happens? It slides slowly across the floor. Give the book a big push. Does it slide much more quickly?

The harder the push, the faster an object travels.

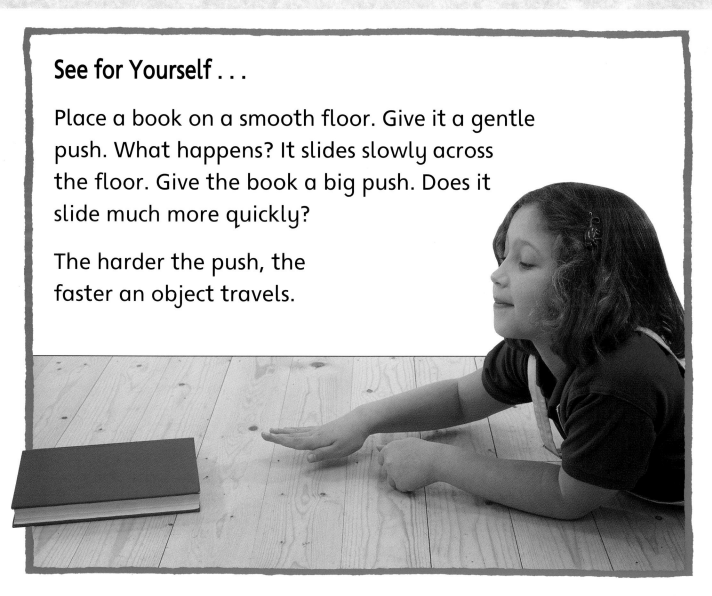

Friction

Friction is a force that slows down moving things. Friction is caused when two objects rub together.

This golf ball is rubbing against the grass. Tiny, rough pieces on both the surface of the ball and the grass push against each other and slow the ball down.

(i) *Friction might stop this ball before it reaches the hole.*

See for Yourself . . .

Place a toy car on the floor. Give it a push. What happens? The friction between the car's wheels and the floor slows the car down. Eventually, the car will stop.

Now put the car on a carpeted floor. The car stops more quickly. Why? Carpet is rougher than smooth wood, so it causes more friction. The car moves for less time.

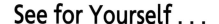

Changing Speed

This ice-skater can choose how quickly she skates and when to stop.

She **controls** her **speed** by the amount of push she uses.

She uses a lot of push to travel quickly but just a little to travel slowly.

To stop, she turns her skates sideways, so that more of the skates rub against the ice.

Can you remember what this rubbing is called?

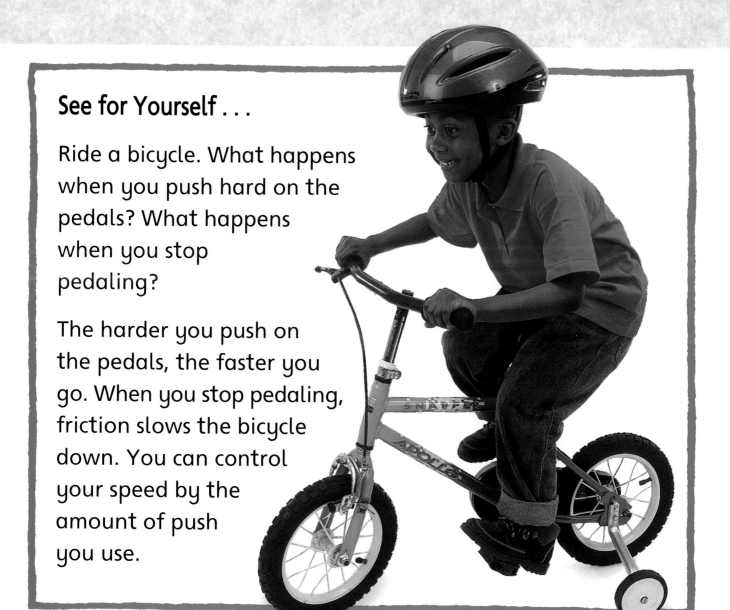

See for Yourself . . .

Ride a bicycle. What happens when you push hard on the pedals? What happens when you stop pedaling?

The harder you push on the pedals, the faster you go. When you stop pedaling, friction slows the bicycle down. You can control your speed by the amount of push you use.

Brakes

This man is riding his bicycle. If he wants to stop, he uses his brakes. When he squeezes the brakes, rubber pads grip the wheels very tightly.

This creates a lot of friction and stops the bicycle very quickly.

? *What other vehicles have brakes?*

See for Yourself . . .

Ask an adult to fasten a paper plate to a stick with a thumbtack as shown.

Spin the plate with your finger.

Hold a straw against the edge of the plate. What happens?

The friction caused by the plate and the straw rubbing together makes the plate stop spinning.

19

Changing Direction

Forces can make things change direction.

Without a force or brakes, once this stroller started moving, it would continue in a straight line. To turn a corner, the baby's father has to make forces act on the stroller.

He pulls back with one hand and pushes forward with the other hand.

 How do car drivers change direction?

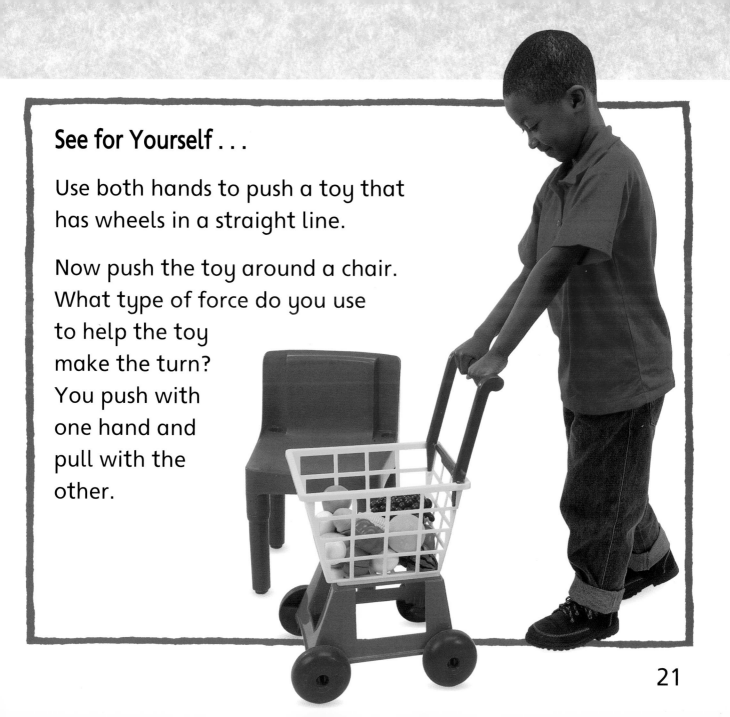

See for Yourself . . .

Use both hands to push a toy that has wheels in a straight line.

Now push the toy around a chair. What type of force do you use to help the toy make the turn? You push with one hand and pull with the other.

21

Changing Shape

Forces can also make things change shape.

The baker in this picture is making bread dough. He is stretching and squashing the dough.

Stretches and squashes are types of forces. They are special pushes and pulls that change the shape of something.

? *What other things can be stretched and squashed?*

See for Yourself . . .

Take some play dough into your hands. Squash and stretch it. Twist it, bend it, and roll it, too. How easy is it to change the dough's shape in many different ways?

Some objects, such as cars, need a large force, like the push of another car in a crash, to change their shapes.

Glossary

Index

Further Readings

Ward, Alan. *Forces and Energy.* F. Watts, 1992.

Peacock, Graham. *Forces.* Thomson Learning, 1994.

Answers

p. 6, your arms; p. 8, automobiles, boats; p. 10, backward; p. 16, friction; p. 18, cars and trains; p. 20, the steering wheel; p. 22, play dough, rubber bands.